FLORA OF TROPICAL EAST AFRICA

TAMARICACEAE

D. R. HUNT

Xerophytic or halophytic shrubs or small trees, with slender branches and reduced subulate or scale-like, alternate, exstipulate, glandular-punctate leaves. Flowers in racemes or spikes or solitary, small, usually bisexual, regular, hypogynous. Calyx and corolla distinct, 5–4-merous, imbricate, free. Disk present. Stamens as many as petals and alternate with them, or twice as many in two alternating whorls; filaments mostly free, sometimes ± united, inserted on the disk; anthers 2-thecous, opening by longitudinal slits. Ovary superior, 1-locular; carpels 3–4; placentas 2–5, basal or parietal, each with 2 or more erect anatropous ovules; styles as many as carpels, free or united below or stigmas sessile. Fruit a capsule. Seeds ∞ with long hairs, with or without endosperm; embryo straight.

A family of four genera, mainly Mediterranean and central Asian in distribution, represented in tropical Africa only by the principal genus.

TAMARIX

L., Sp. Pl.: 270 (1753) & Gen. Pl., ed. 5: 131 (1754)

Flowers in narrow racemes or spikes. Petals without ligules. Stamens free or shortly connate at the base; anthers extrorse. Ovary narrowed towards the apex; placentas basal; styles short and thick; stigmas flattened or somewhat concave. Seeds with a sessile apical tuft of hairs; endosperm absent.

A genus widespread around the Mediterranean, in Europe and central Asia, reaching South Africa, India, China and Japan. No reliable estimate of the number of valid species can be given now, though it would probably not exceed fifty. A number of species are popular as ornamentals.

T. gallica L. of European origin has been introduced into cultivation on the banks of Lake Naivasha, Kenya (*W. May* in *E.A.H.* H164/56–7 !). It is distinguished from *T. nilotica* (Ehrenb.) Bunge by the purplish-brown larger branchlets (usually light brown in *T. nilotica*), the short pink racemes, the shorter floral bracts not exceeding the ovate acute sepals, and the filaments broadened at the base.

Smaller branchlets apparently articulated; leaves reduced
to a minutely apiculate sheathing base . . . 1. *T. aphylla*
Smaller branchlets not articulated; leaves imbricated, half-
clasping, 1–3 mm. long . . . . . . 2. *T. nilotica*

1. **T. aphylla** (*L.*) *Karst.*, Deutsch Fl.: 641 (1880–3); Lanza in Bull. Ort. Bot. Palermo 8: 82 (1909); D. R. Hunt in K.B. 16: 481 (1963). Type: central Asia, *Herb. Linnaeus* 1136.3 (LINN, lecto. !)

Tree up to 10(–15) m. high. Smaller branchlets apparently articulated, enclosed by the abruptly truncate and minutely apiculate sheathing leaf-bases 1·5–4 mm. long, 1 mm. in diameter, usually with a dusty grey glandular efflorescence; larger branchlets with persistent scale-like leaves. Racemes crowded near end of current year's growth, usually 4–6 cm. long, 3–4 mm. in

D.E.

FIG. 1. *TAMARIX NILOTICA*—**1**, leafy branch, × 1; **2**, portion of branchlet, × 5; **3**, flowering shoot, × 1; **4**, detail of inflorescence, × 4; **5**, side view of flower with one petal removed, × 10; **6**, stamens and pistil, × 10; **7**, stamens and disk (flattened) from above, × 10; **8**, ovary with side removed to show placentation, × 12; **9**, ripe fruit, side view, × 10; **10**, seed, × 10. *T. APHYLLA*—**11**, portion of branchlet × 4; **12**, detail of inflorescence, × 4. 1–4, from *Bogdan* 2325; 5–8, from *Battiscombe* 38; 9, from *Greenway* 3943; 10, from *Gillett* 13301; 11, 12, from *J. Adamson* 83.

diameter; bracts sheathing, concave, less than 1 mm. long, acute. Flowers subsessile, about 3 mm. long, 1·5 mm. wide. Sepals 5, suborbicular, 1 mm. long, greenish or pinkish. Petals 5, oblong, 2 mm. long, 1 mm. wide, obtuse, white. Disk irregularly lobed between insertion of filaments, dull red. Stamens 5; filaments equalling or exceeding petals, slender, white; anthers shortly apiculate. Ovary about 1·5 mm. long; carpels 3; styles erect, suberect or incurved, one-third to one-half as long as ovary; stigmas elliptic, concave. Valves of ripe capsule 4–5 mm. long. Seeds terete, 0·5 mm. long, brown; apical coma 3 mm. long. Fig. 1/11,12.

KENYA. Northern Frontier Province: between Yabichu and Mandera, 23 May 1952, *Gillett* 13301! & between Mandera and Ramu, 10 Oct. 1955, *J. Adamson* 83!
DISTR. **K**1; Morocco, Algeria, Libya, Egypt, Sudan Republic, Eritrea, Ethiopia and Somali Republic; also in Israel, Persia, Afghanistan and India
HAB. Beside seasonal rivers in deciduous bushland; 200–400 m.

SYN. *Thuja aphylla* L., Cent. Pl. 1: 32 (1755), pro parte, excl. syn. *Cupressus fructu quadrivalvi*...Shaw. Afr.: 188. (See Franco in Bol. Soc. Brot., sér. 2, 25: 219 (1951) & D. R. Hunt in K.B. 16: 481 (1963))
   Tamarix orientalis Forsk., Fl. Aegypt.-Arab.: 206 (1775). Type: Egypt, *Forskål* (C, holo.!)
   T. articulata Vahl, Symb. Bot. 2: 48, t. 32 (1791); Oliv. in F.T.A. 1: 151 (1868), *nom. illegit.* Type: as for *T. orientalis*

2. **T. nilotica** (*Ehrenb.*) *Bunge*, Tent. Gen. Tamaricum: 54 (1852); T.S.K.: 16 (1936); T.T.C.L.: 605 (1949); K.T.S.: 552 (1961). Type: Egypt, *Ehrenberg* (K, ?iso.!)

Shrub or small tree up to 6 m. high. Leaves sessile, narrowly lanceolate to narrowly ovate, concave, closely imbricated on and half-clasping the smaller branchlets, 1–3 mm. long (4–5 mm. on larger branchlets), acute to long-acuminate. Racemes crowded near end of current year's growth, rather lax, 5–9 cm. long, 3–5 mm. in diameter; bracts narrowly lanceolate 2(–3) mm. long, exceeding flowers (in East African specimens). Flowers 1·75 mm. long, 1·25 mm. wide, white, on 0·5 mm. long pedicels. Sepals 5, ovate to rotund, about 1 mm. long, entire or minutely denticulate. Petals 5, oblong, 1·5–2 mm. long, 0·75–1 mm. wide, obtuse, falling early. Stamens 5(–7); filaments finally 2 mm. long, slender; anthers shortly apiculate. Disk truncate or slightly lobed between stamens. Ovary pyriform, 0·75–1 mm. long; carpels 3; styles erect or suberect; stigmas minutely lobed. Valves of ripe capsule 4–5 mm. long. Seeds terete, 0·5 mm. long, brown; apical coma 1·5–2·5 mm. long. Fig. 1/1–10.

KENYA. Turkana District: SW. Lake Rudolf, *Champion* T45!; Machakos District: Lesser Kiboko R., 18 Feb. 1949, *Bogdan* 2325!; Tana River District: Garissa, 26 Sept. 1957, *Greenway* 9232!
TANGANYIKA. Shinyanga District: Mwadui Diamond Mine compound, 10 Mar. 1964, *Carmichael* 1064!; Lushoto District: Mkomazi, 22 Apr. 1934, *Greenway* 3943! & July 1955, *Semsei* 2147!; Singida District: shores of Lake Singida, 17 July 1936, *H. Fraser* 975!
DISTR. **K**2, 4, 7; **T**1–3, 5; Egypt, Sudan Republic, Ethiopia, Somali Republic; also in Israel and Arabia
HAB. Banks of seasonal rivers and shores of lakes, occurring in vegetation varying from deciduous woodland to semi-desert scrub; 180–1340 m.

SYN. *T.* (*gallica*) *nilotica* Ehrenb. in Linnaea 2: 269 (1827). (Given in this form by Ehrenberg with the status of " Hauptvarietät ")
   [*T. gallica* sensu Oliv. in F.T.A. 1: 151 (1868), pro parte, *non* L. sensu stricto]

NOTE. The specific status of this taxon is accepted largely for practical reasons. The genus *Tamarix*, and especially the *T. gallica* complex, is particularly intricate, and a monographic revision is badly needed. Until such a treatment is available, no useful purpose is served by tampering with the taxonomy and substituting an unwieldy or less apt name. It is as well to point out, however, that only minor differences separate *T. nilotica* from *T. indica* Willd. (*T. senegalensis* DC.), and that both species may be more correctly interpreted as geographical subspecies of *T. gallica* L.

INDEX TO TAMARICACEAE